TASTING

by Robin Nelson

first step nonfiction

Lerner Publications Company · Minneapolis

Tasting is one of my **senses.**

I taste with my tongue.

I taste something **sweet.**
I taste honey.

I taste an apple.

I taste something **sour.**
I taste a lemon.

I taste a pickle.

I taste something **salty.**
I taste french fries.

I taste a pretzel.

I taste something **bitter.**
I taste coffee.

I taste medicine.

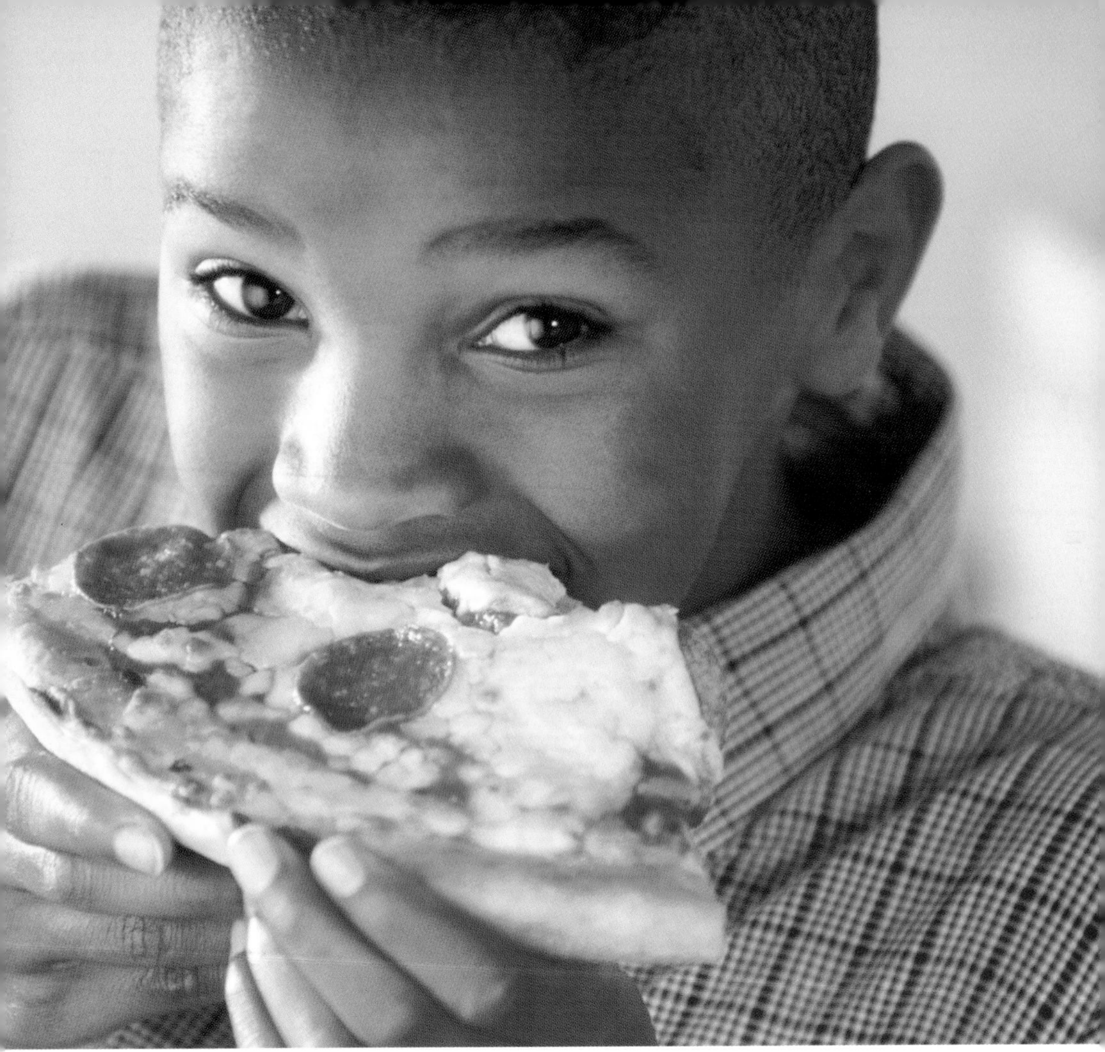

I taste something I like.
I taste pizza.

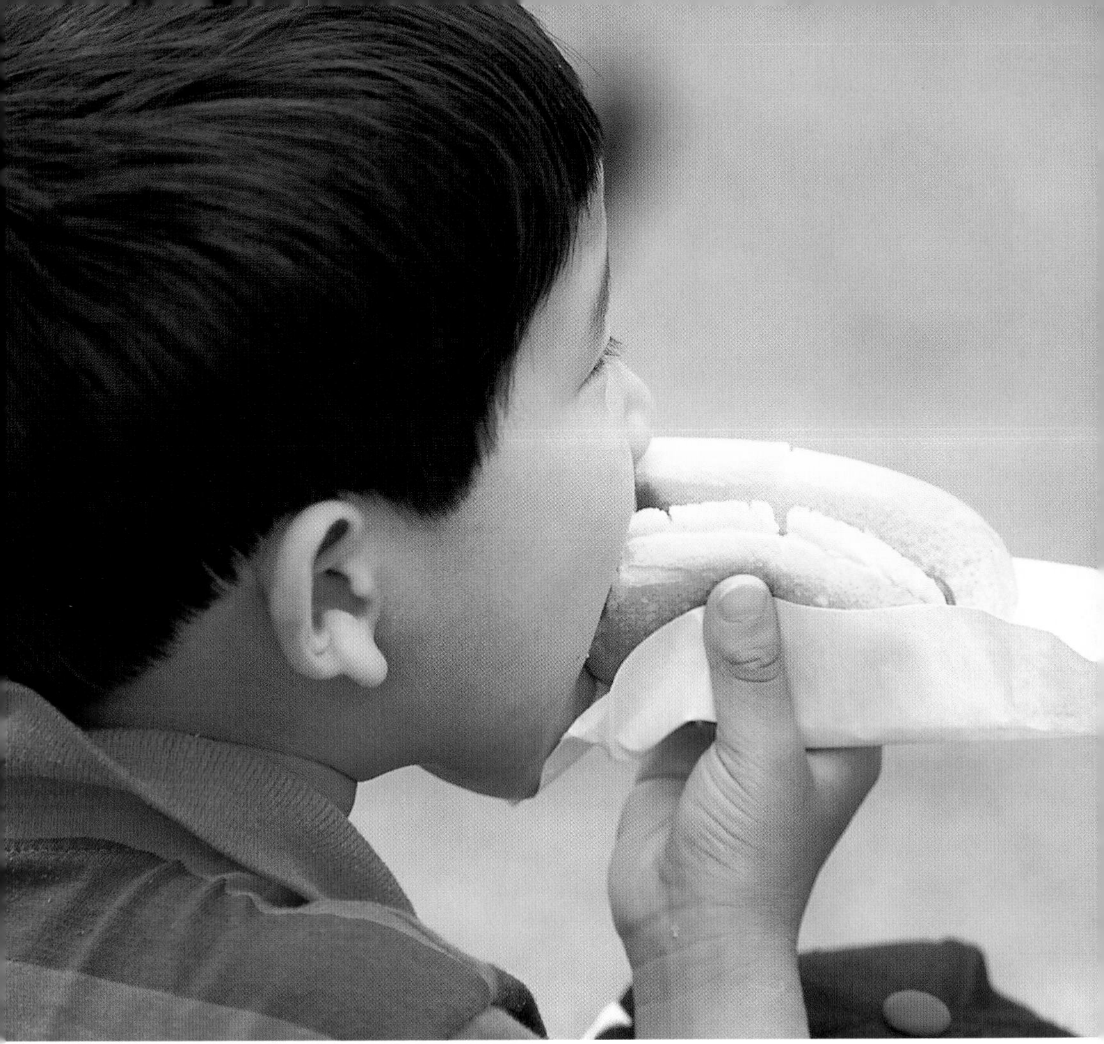

I taste a hot dog.

I taste something I don't like. I taste fish.

I taste peas.

I taste many things.

What do you taste?

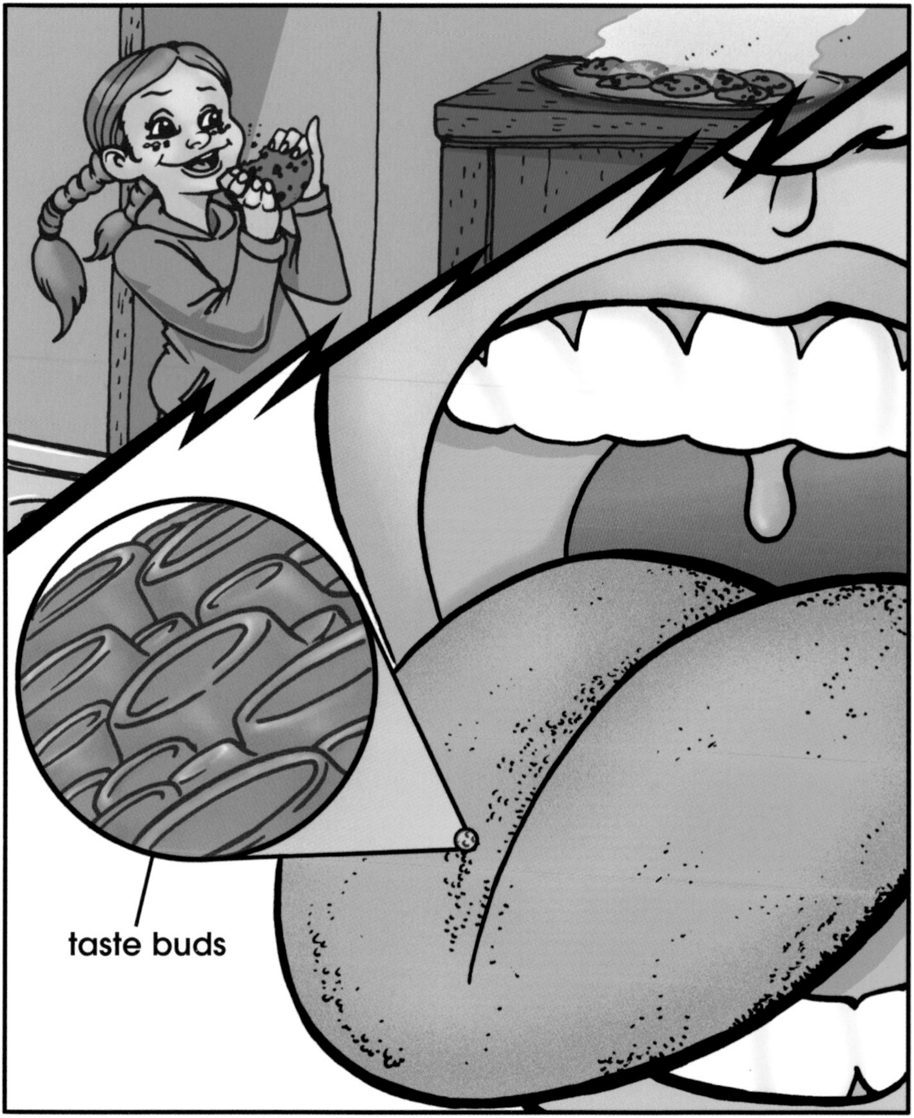
taste buds

How do you taste?

You taste things with your tongue. There are hundreds of tiny taste buds all over your tongue. When you eat something, the taste buds detect different kinds of tastes—bitter, salty, sweet, or sour.

Tasting Facts

- The sense of taste is found in your taste buds on your tongue.

- Insects have the best sense of taste. They have taste organs on their feet, antennae, and mouth parts.

- Birds will not eat monarch butterflies because the butterflies taste bad.

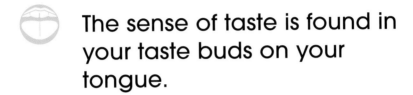

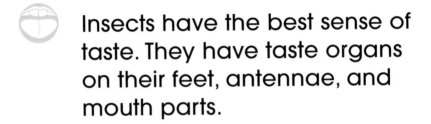

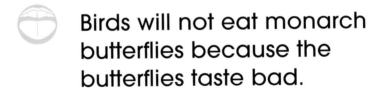

- Butterflies have taste buds on their feet. They walk on a flower to decide if it is good enough to eat.

- Fish can taste with their fins and tail as well as their mouth.

- In general, girls have more taste buds than boys do.

- Taste is the weakest of the five senses.

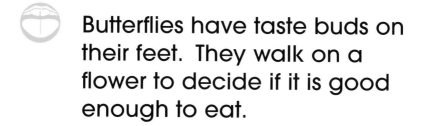

Glossary

bitter – having a sharp taste

salty – having a taste like salt

senses – the five ways our bodies get information. The five senses are hearing, seeing, smelling, tasting, and touching.

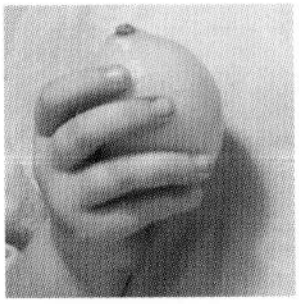

sour – having an acid taste

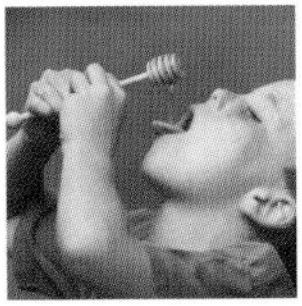

sweet – having a taste like sugar

Index

Cover image used courtesy of: © RubberBall Royalty Free Digital Stock Photography.

Photos reproduced with the permission of: © RubberBall Royalty Free Digital Stock Photography, pp. 2, 17, 22 (middle); © Amos Nachoum/CORBIS, p. 3; © Lynda Richardson/CORBIS, pp. 4, 22 (bottom); Brand X Pictures, p. 5; © Richard B. Levine, pp. 6, 22 (second from top); © Todd Strand/Independent Picture Service, p. 7; © Kevin Fleming/CORBIS, pp. 8, 22 (second from top); © Karen Huntt Mason/CORBIS, p. 9; © Annie Griffiths Belt/CORBIS, pp. 10, 22 (top); © 2001 PhotoDisc, pp. 11, 12; © Phyllis Picardi/Photo Network, p. 13; © Owen Franken/CORBIS, p. 14; © 2001 Mitch York/Stone, p. 15; © Layne Kennedy/CORBIS, p. 16.

Illustration on page 18 by Tim Seeley.

Lerner Publications Company
A division of Lerner Publishing Group
241 First Avenue North
Minneapolis, MN 55401 U.S.A.

Website address: www.lernerbooks.com

Library of Congress Cataloging-in-Publication Data

Nelson, Robin, 1971–
Tasting / by Robin Nelson.
p. cm. — (First step nonfiction)
Includes index.
Summary: An introduction to the sense of taste and the different things that you can taste.
ISBN: 0-8225-1265-3 (lib. bdg. : alk. paper)
1. Taste—Juvenile literature. [1. Taste. 2. Senses and sensation.]
I. Title. II. Series.
QP456 .N45 2002
612.8'7—dc21 2001003962

Manufactured in the United States of America
2 3 4 5 6 7 – DP – 09 08 07 06 05 04